Process Intensification 2019

Topical Conference at the 2019 AIChE Spring Meeting
and 15th Global Congress on Process Safety

New Orleans, Louisiana, USA
31 March - 4 April 2019

ISBN: 978-1-5108-8684-1

Printed from e-media with permission by:

Curran Associates, Inc.
57 Morehouse Lane
Red Hook, NY 12571

Some format issues inherent in the e-media version may also appear in this print version.

Printed by Curran Associates, Inc. (2019)

AIChE
120 Wall Street, FL 23
New York, NY 10005-4020

Phone: (800) 242-4363
Fax: (203) 775-5177

www.aiche.org

Additional copies of this publication are available from:

Curran Associates, Inc.
57 Morehouse Lane
Red Hook, NY 12571 USA
Phone: 845-758-0400
Fax: 845-758-2633
Email: curran@proceedings.com
Web: www.proceedings.com

TABLE OF CONTENTS

549343 Ammonia Manufacturing Box: *Small Ammonia Plant Design Using Micro Processing Technology and Process Intensification Technology*

Ammonia Manufacturing Box

Small Ammonia Plant Design using Micro Processing Technology and Process Intensification Technology

By: Richard B. Strait, Adjunct Professor, Rice University, Department of Chemical and Biomolecular Engineering, Houston, Texas

The small self-contained ammonia manufacturing box (AMB) came out of the need for farmers in remote locations containing modest amounts of natural gas and water to produce enough fertilizer for their local needs.

Presently ammonia plants are getting larger and larger. The future however, is to make small highly automated plants that are portable, use readily available inputs and produce the products where they are needed. To this end the new small scale plant will include all the latest leading edge technologies that apply such as Micro Processing Technology (MPT), Process Intensification (PI) and automated safety systems. The ammonia plant of the future is contained on one rail car, shop fabricated and contains all the latest technology.

Recent developments in micro reactor designs are applied to the ammonia plant reactions; with special attention to: A) steam methane reforming, B) water gas shift and C) ammonia synthesis.

1. Steam methane reforming occurs at high temperatures, is endothermic and an equilibrium reaction. Alternatives are evaluated including: 1) electric reformer, 2) fired heater reformer and 3) auto thermal reformer.
2. Water gas shift occurs at high temperatures with one catalyst for optimum rate and at lower temperatures with a different catalyst to reach a more complete conversion for this endothermic equilibrium reaction. Alternatives are evaluated including: 1) two shift reactors with precise temperature control along the flow path and 2) one reactor with a changing catalyst formulation along the reactor tube length.
3. Ammonia synthesis occurs at high pressures and high temperatures, is exothermic and an equilibrium reaction. Alternatives are evaluated including: 1) precise reactor temperature control following the equilibrium curve and 2) micro channel liquid separation of product ammonia.

Process intensification is applied to CO2 removal, compression, refrigeration and heat integration as well as alternative energy sources.

Safety, operability and maintenance are improved over the traditional stick built plant and these important issues are addressed.

R. B. Strait

October 24, 2018

550545 A Comparison of the Decompositions of Benzene and Toluene As Tar Representatives Using Non-Thermal Plasma

Non-thermal plasma is being considered as an alternative technology to decompose gasification tars. In this study, a dielectric barrier discharge (DBD) reactor was used to remove the tar representatives (benzene and toluene) in a model product gas (CH_4, CO_2, CO and H_2). The effect of specific input energy (SIE) (2.05-16 kWh/m^3), residence time (0.95-2.82s), and temperature (20-400°C) were studied. The decomposition of benzene and toluene increased with increasing SIE and residence time. At ambient temperatures, the formation of lower hydrocarbons, and solid residue occurred. This would be a significant operational problem, but here we show that the effect can be significantly reduced by operation at higher temperatures. In the presence of plasma at 400 °C, solid formation was substantially reduced. At the same time, selectivity to lower hydrocarbons increased from 31 to 74 % and 36 to 79 % for toluene and benzene, respectively. However, the decomposition of CO_2 and CH_4 decreased with increasing the temperature.

550506 Grams - a Dynamic Intensification and Optimization Platform for Modular Chemical Process Systems

A significant portion of future demands for chemical feedstocks is predicted to be met by unconventional and distributed resources such as associated shale gas, stranded natural gas, coalbed methane, landfill gas, biogas, and fuel gas. Current roadblocks to widespread utilization of these resources often include a lack of pipeline infrastructure, sparsity, seasonal variability and uncertainty in feedstock volumes and compositions, and occasional high levels of contamination. Small-scale, modular chemical process systems (MCPS) can be a promising solution to tap into these unconventional resources. They offer increased flexibility, agility and ability to accommodate varying market demands. However, poor economies-of-scale and high capital intensity prohibit most small-scale MCPS to be economically viable over their large-scale counterparts. We argue that this limitation can be overcome through dynamic process intensification that combines multiple phenomena, such as separation, conversion and intermittent storage, within a single equipment in a periodic manner. One example of such a multifunctional MCPS technology is the sorption-enhanced reaction process (SERP). In SERP systems, *in situ* capture of undesired reaction byproducts by adsorbents favors the desired reactions in catalysts, thereby increasing the overall yield of desired products. To this end, advanced modeling, simulation and optimization of such dynamically intensified and complex systems can provide critical breakthroughs needed for small-scale MCPS towards the economic production of chemicals utilizing unconventional feedstocks.

We have developed GRAMS (Generalized Reaction-Adsorption modeling, optimization and Simulation), which is a first-of-its-kind computational framework for the optimal design and operation of dynamically intensified packed-bed systems. For given feeds and products specifications, GRAMS is used to optimally design the cycle configurations, column design specifications and process operating conditions of a periodically operated MCPS [1-3]. The GRAMS platform combines an in-house dynamic process simulator with a cost estimator and a data-driven constrained grey-box optimizer [4]. The high-fidelity process simulator is based on a first-principles-based model of a generalized reaction-adsorption system to predict the performances of processes incorporating periodic gas adsorption-desorption (e.g., pressure/temperature/vacuum swing adsorption: PSA, TSA, VSA, SMB), reaction (e.g., PFR, SMBR), or a combination of both (e.g., SERP). The model predictions are extensively validated with experimental data for industrially relevant pressure swing adsorption (PSA), steam methane reforming (SMR), methanol synthesis, sorption-enhanced SMR (SE-SMR), and sorption-enhanced water gas shift reaction (SE-WGSR) processes. The framework has been used for the optimal synthesis of three multi-mode, multi-step and periodic SERP systems, namely SE-SMR, SE-WGSR and sorption-enhanced methanol (SE-MeOH). The optimized SE-SMR produces hydrogen from natural gas with 35% higher productivity and more than 10% lower cost in comparison to existing small-scale systems [2]. Furthermore, the novel SE-MeOH process, designed using GRAMS, could lead to more than 7% improvement in methanol yield with only 2% decrease in production capacity [3].

References

[1] A. Arora, S. S. Iyer, and M. M. F. Hasan, "GRAMS: A General Framework Describing Adsorption, Reaction and Sorption-Enhanced Reaction Processes," *Chem. Eng. Sci.*, vol. 192, pp. 335–358, 2018.

[2] A. Arora, I. Bajaj, S. S. Iyer, and M. M. F. Hasan, "Optimal Synthesis of Periodic Sorption Enhanced Reaction Processes with Application to Hydrogen Production," *Comput. Chem. Eng.*, vol. 115, pp. 89–111, 2018.

[3] A. Arora, S. S. Iyer, I. Bajaj, and M. M. F. Hasan, "Optimal Methanol Production via Sorption Enhanced Reaction Process," *Ind. Eng. Chem. Res.*, 2018.

[4] I. Bajaj, S. S. Iyer, and M. M. F. Hasan, "A Trust Region-based Two Phase Algorithm for Constrained Black-box and Grey-box Optimization with Infeasible Initial Point," *Comput. Chem. Eng.*, 2017.

550635 Low-Temperature Fischer – Tropsch Synthesis Multiphase Product Recovery through Reactive Distillation

The Fischer–Tropsch (FT) process relies on a catalyzed chemical reaction to convert synthesis gas into paraffins, olefins, and an appreciable amount of oxygenates. The synthesized FT product slate is remarkable, ranging from C_1 (methane) up to C_{70+}. The distribution of FT products varies based on reaction conditions (i.e., temperature, pressure, space velocity, and H2/CO ratio), reactor type, and catalyst formulation. We have developed an equation-oriented framework for optimal synthesis of integrated reactive distillation (RD) systems for FT processes in GAMS in which the phase equilibrium is given by a cubic equation of state, the reaction rate expressions are expressed in terms of fugacities, and the product selectivity of catalysts are based on the Anderson–Schulz–Flory distribution and experimental data [1].

In this work, we are expanding the model to consider a more comprehensive kinetic expression with a wider product selectivity to explore the potential of producing a predominantly wax product on a low-temperature FT process, with an expanded model to optimize tray numbers, feed tray location and catalyst loading on trays. We are also implementing the FT-RD optimization model within the PyOMO optimization framework.

The results show that exothermic heat of reaction is sufficient to provide enough vapor load which renders a reboiler unnecessary; and side-streams of high quality gasoline and diesel can be obtained. Thus demonstrating that novel reactive distillation processes are paving the way for a more sustainable chemical process industry that is profitable, safer and less polluting.

[1] Y. Zhang, C.M. Masuku, L.T. Biegler, Equation-oriented framework for optimal synthesis of integrated reactive distillation systems for the Fischer–Tropsch processes, Energy & Fuels 32 (2018) 7199–7209.

550159 Towards the Incorporation of Operability and Safety in the Synthesis of Intensified Reactive and Extractive Separation Systems

Intensified processes typically feature higher degree of integration of multifunctional phenomena, less degrees of freedom, narrower operability windows, and even faster dynamics [1-2]. Thus, the incorporation of control, operability, and safety assessment at the design level becomes even more critical to ensure the actual operational performances in these systems, comparing to their conventional unit-operation-based counterparts. Despite recent advances [3-6], systematic synthesis approaches and analytical tools for the design and operability/safety optimization of intensification processes are still lacking.

To address this challenge, in this work and as part of the RAPID SYNOPSIS Project [7], we propose a systematic framework for the synthesis of operable process intensification systems, with specific focus on reactive and extractive separation processes. This framework is based on a phenomenological process intensification/synthesis approach (i.e., Generalized Modular Representation Framework (GMF) [8-10]), which first identifies candidate/promising intensified tasks and then translates them to equipment-based flowsheet alternatives. Flexibility analysis is integrated with the GMF model to ensure that resulting design configurations can be operable under varying operating conditions. To systematically account for inherent safety performance, risk assessment criteria are included as process constraints considering failure frequency and consequence severity. Simultaneous design and control strategies are also employed to ensure validated operable intensified design configurations using advanced multi-parametric model-based predictive control (mp-MPC) developed via the PAROC (i.e., PARametric Optimization and Control) framework [11]. Two case studies are presented to highlight the potential of the proposed approach: (i) a reactive separation system - methyl-tert-butyl-ether production, and (ii) an extractive separation system - methanol/water azeotropic separation.

References

1. Tian, Y., Demirel, S. E., Hasan, M. F., & Pistikopoulos, E. N. (2018). An Overview of Process Systems Engineering Approaches for Process Intensification: State of the Art. Chemical Engineering and Processing-Process Intensification, 133, 160-210.

2. Nikačević, N. M., Huesman, A. E., Van den Hof, P. M., & Stankiewicz, A. I. (2012). Opportunities and challenges for process control in process intensification. Chemical Engineering and Processing: Process Intensification, 52, 1-15.

3. Baldea, M. (2015). From process integration to process intensification. Computers & Chemical Engineering, 81, 104-114.

4. Carrasco, J. C., & Lima, F. V. (2018). Bilevel and parallel programing based operability approaches for process intensification and modularity. AIChE Journal, 64(8), 3042-3054.

5. Nemet, A., Lee, K., Klemeš, J. J., Varbanov, P. S., Moon, I., & Kravanja, Z. (2015). Safety analysis embedded in total site synthesis. In Chemical Engineering Transactions. Italian Association of Chemical Engineering-AIDIC.

6. Medina-Herrera, N., Grossmann, I. E., Mannan, M. S., & Jiménez-Gutiérrez, A. (2014). An approach for solvent selection in extractive distillation systems including safety considerations. Industrial & Engineering Chemistry Research, 53(30), 12023-12031.

7. SYNOPSIS - Synthesis of operable process intensification systems (DE-EE0007888-09-03).

8. Papalexandri, K. P., & Pistikopoulos, E. N. (1996). Generalized modular representation framework for process synthesis. AIChE Journal, 42(4), 1010-1032.

9. Ismail, S. R., Proios, P., & Pistikopoulos, E. N. (2001). Modular synthesis framework for combined separation/reaction systems. AIChE Journal, 47(3), 629-649.

10. Tian, Y., Mannan, M. S., & Pistikopoulos, E. N. (2018). Towards a systematic framework for the synthesis of operable process intensification systems. In Computer Aided Chemical Engineering (Vol. 44, pp. 2383-2388). Elsevier.

11. Pistikopoulos, E. N., Diangelakis, N. A., Oberdieck, R., Papathanasiou, M. M., Nascu, I., & Sun, M. (2015). PAROC—An integrated framework and software platform for the optimisation and advanced model-based control of process systems. Chemical Engineering Science, 136, 115-138.

550636 Building Block-Based Design and Intensification of Chemical Processes

Process intensification provides means for more sustainable, cost effective, smaller and safer designs compared to their conventional counterparts [1-2]. Often times, intensification is realized through capturing the synergy between different process phenomena. Identification of such intensification opportunities can be facilitated through a departure from unit-operation-based design paradigm which has been the modus operandi since the last century. Accordingly, several process intensification methods have been developed in the past that rely on process phenomena, mass/heat exchange modules and/or tasks [3-4]. However, it remains a challenge to systematically identify the optimal intensified flowsheets for a given design problem. To this end, our recently proposed building block-based approach [5] provides unique advantages for systematic process design and intensification [6-8]. In this work, we will show that this representation approach leads to an optimization-based systematic method for process intensification. Unlike the traditional superstructure-based approaches, our proposed building block superstructure relies on physicochemical phenomena to automatically generate optimal intensified flowsheets. It is formed by building blocks positioned on a two-dimensional grid and connectivity between the blocks is obtained through inter-block mass and energy transfer. By using either single or multiple building blocks, many physicochemical phenomena can be represented. This results in a mixed-integer nonlinear optimization (MINLP)-based model that describes the superstructure. We will demonstrate the benefits of the building block-based design approach in terms of systematic process intensification through several case studies. Special focus will be given to intensification of natural gas utilization systems, energy-intensive separations, and reactive separation processes.

[1] Stankiewicz, A. I., Moulijn, J. A. (2000). Process Intensification: Transforming Chemical Engineering. Chemical Engineering Progress 1, 22–34.

[2] Tian, Y., Demirel, S. E., Hasan, M. M. F., Pistikopoulos, E. N. (2018). An Overview of Process Systems Engineering Approaches for Process Intensification: State of the Art. *Chemical Engineering and Processing: Process Intensification*, 133, 160-210.

[3] Lutze, P., Babi, D.K., Woodley, J.M. and Gani, R. (2013). Phenomena based methodology for process synthesis incorporating process intensification. *Industrial & Engineering Chemistry Research*, *52*(22), pp.7127-7144.

[4] Papalexandri, K. P., Pistikopoulos, E. N. (1996). Generalized modular representation framework for process synthesis. *AIChE Journal*, *42*(4), 1010-1032.

[5] Demirel, S. E., Li, J., and Hasan, M. M. F., (2017). Systematic Process Intensification using Building Blocks, *Computers and Chemical Engineering*, 105, 2-38.

[6] Li, J., Demirel, S.E. and Hasan, M.M.F., 2018. Process synthesis using block superstructure with automated flowsheet generation and optimization. AIChE Journal, 64(8), 3082-3100.

[7] Li J., Demirel S.E., Hasan M.M.F. Process Integration using Block Superstructure. *Industrial & Engineering Chemistry Research*, 2018, 57: 4377–4398.

[8] Li J., Demirel S.E., Hasan M.M.F. Fuel Gas Network Synthesis Using Block Superstructure. *Processes*. 2018, 6(3): 23.

549876 Semi-Continuous Processing to Improve Safety and Quality

Challenges exist for processing viscous products that are classified as flammable due to high solvent content. This paper will focus on the processing options for mixing a highly filled shear thinning base material with a low viscosity flammable solvent followed by direct filling into final packaging.

This process can be accomplished using a change can batch mixer, but with significant drawbacks. Injecting hazardous and/or flammable materials through a hand held cartridge device into a pressurized vessel, such as a change can mixer, has potential for operator exposure and relies on personal protective equipment for protection. Another drawback is that batch processing results in a large quantity of work in progress (WIP) material with potential for failure of an entire batch leading to high scrap costs. To avoid these problems an alternate process was sought out to allow for in process adjustments to hit quality targets while minimizing scrap and improving safety.

Multiple solutions were considered. Use of a continuous mixing unit such as a twin screw extruder or pin mixer was examined. Although the mixing in these units was sufficient and provided higher processing rates, material properties were not easily adjusted. A semi-continuous option utilizing a series of static mixers fed through volume controlled feed units with direct filling into the final packaging was then explored. Several potential hurdles were overcome through multiple trials demonstrating the ability to use this process scheme to achieve our processing goals.

By using the semi-continuous process scheme it was possible to reduce the WIP material by 70% compared to the batch change can mixer. Quality was improved by the ability for in-process testing and process productivity increased. Potential chemical exposure for operators was reduced through an upgraded design of the feed systems.

550628 Synthesis of Reactive Separation Systems Via Building Blocks

Process intensification (PI) has become an important tool for novel process designs featuring drastic improvements in process performance. While miniaturization, use of alternative energy sources and discovery of multifunctional materials are examples for activities leading to intensification, integration of several operations in a single equipment have been one of the most successful PI strategies in terms of their industrial acceptance [1-2]. Specifically, reactive distillation (RD) units, which combine reaction and distillation operations in a single column, have seen a marked increase in their industrial adoption since 1980s [2]. Wider utilization of this technology requires systematic design procedures suggesting optimal intensified designs at the early design stage. Several methods have been suggested in the past for the synthesis of reactive distillation operations [3-4]. In this work, we use building block-based representation [5-8] to provide a systematic procedure for the synthesis and optimization of reactive separation processes. This novel representation can suggest non-intuitive intensified configurations with its unique outlook towards process design. The representation is based on collection of blocks positioned on a two-dimensional grid which can be used to represent several physical and chemical phenomena. Each building block has several physical attributes, i.e. temperature, pressure, phase, composition, and can stand for a unit use of a material, e.g. catalyst. While reaction operations can be represented within a single block, separation related phenomena require at least two blocks with different phases. Vapor-liquid equilibrium, for instance, requires one vapor and one liquid block sharing a common boundary through which phase transition can take place. When reaction operation is assigned into one of these neighboring blocks, a reactive vapor-liquid contact stage can be obtained. Collection of these building blocks in a grid formation yields building block superstructure which is modeled with a Mixed Integer Non-linear Programming (MINLP) model. Continuous variables are used to quantify the material and energy flow between the blocks and integer variables are utilized to assign different phenomena at different positions on the grid. This results in a generic representation and synthesis method for reactive separation processes. We will demonstrate the applicability of the proposed method with several example problems including ideal and non-ideal reactive systems and show that cost-optimal intensified designs can be realized through building-block based approach.

[1] Stankiewicz, A. I., Moulijn, J. A. (2000). Process Intensification: Transforming Chemical Engineering. Chemical Engineering Progress 1, 22–34.

[2] Tian, Y., Demirel, S. E., Hasan, M. M. F., Pistikopoulos, E. N. (2018). An Overview of Process Systems Engineering Approaches for Process Intensification: State of the Art. Chemical Engineering and Processing: Process Intensification, 133, 160-210.

[3] Ismail, S.R., Pistikopoulos, E.N. and Papalexandri, K.P. (1999). Synthesis of reactive and combined reactor/separation systems utilizing a mass/heat exchange transfer module. *Chemical engineering science*, *54*(13-14), 2721-2729.

[4] Okasinski, M.J. and Doherty, M.F. (1998). Design method for kinetically controlled, staged reactive distillation columns. *Industrial & Engineering Chemistry Research*, *37*(7), 2821-2834.

[5] Demirel, S. E., Li, J., and Hasan, M.M.F. (2017). Systematic Process Intensification using Building Blocks, *Computers and Chemical Engineering*, 105, 2-38.

[6] Li, J., Demirel, S.E. and Hasan, M.M.F. (2018). Process synthesis using block superstructure with automated flowsheet generation and optimization. *AIChE Journal*, 64(8), 3082-3100.

[7] Li J., Demirel S.E., Hasan M.M.F. (2018). Process Integration using Block Superstructure. *Industrial & Engineering Chemistry Research*, 57, 4377–4398.

[8] Li J., Demirel S.E., Hasan M.M.F. (2018). Fuel Gas Network Synthesis Using Block Superstructure. *Processes*. 2018, 6(3), 23.

550417 A Comparative Study of Residence Time Distribution and Cooling Crystallization in a Continuous Dynamic/Oscillatory Baffle Crystallizer Versus a Stirred Tank

Crystallization has become a predominant technique in particle design technology and it is widely applied in the pharmaceutical industry. Traditionally crystallization has been carried out in batch mode where operating conditions vary with time, often resulting in inconsistent products.[1] Continuous crystallization, on the other hand, has been identified to improve reproducibility, reduce operating cost and simplify scale-up.[2] A novel commercial dynamic baffle crystallizer(DBC) operated as a mixed-suspension-mixed-product-removal(MSMPR) system has been studied. It consists of a reactor tank and dynamic 'donut' shaped baffles to provide oscillatory flow resulting in more uniform mixing.[3] Oscillatory systems have been well studied for batch reactions and synthesis but few studies have been carried out to study the DBC for continuous two-phase operations such as crystallization.

In the work proposed here, liquid and solid residence time distributions were studied systematically in the DBC in comparison with a traditional continuous stirred tank crystallizer(STC). In-situ process analytical technology(PAT) tools such as UV-vis, Infrared(IR), and focused beam reflectance measurement(FBRM) were used to measure system response to step or pulse changes in order to construct RTDs. It is observed that the uniformity of both liquid and solid RTDs was significantly improved in the DBC than the STC indicating potential improvement in size uniformity during crystallization processes. A preliminary continuous cooling crystallization of paracetamol was carried out for two different oscillating conditions in the DBC as well as the STC at the corresponding rotational speeds yielding (approximately) equal power density. Online PAT tools were used to monitor the processes in addition to sampling for offline characterization at every residence time (45min). Nucleation event took place earlier in the DBC than the STC possibly due to the improved mixing mechanism. At early stages of operation, DBC produced crystalline particles of comparable or improved size uniformity. At later stages of operation (>7 residence times), particle sedimentation or severe particle breakage occurred in the STC causing process deviation from steady state while such events did not take place in the DBC and steady state was maintained. In the RTD studies as well as the preliminary crystallization study, the DBC showed improved system uniformity and consistency demonstrating potentials for more in-depth crystallization studies in the DBC.

1. Nagy ZK, Aamir E. Systematic design of supersaturation controlled crystallization processes for shaping the crystal size distribution using an analytical estimator. *Chem Eng Sci*. 2012;84:656-670. doi:10.1016/j.ces.2012.08.048.
2. Alvarez AJ, Singh A, Myerson AS. Crystallization of Cyclosporine in a Multistage Continuous MSMPR Crystallizer. *Cryst Growth Des*. 2011;11(10):4392-4400. doi:10.1021/cg200546g.
3. Hewgill MR, Mackley MR, Pandit AB, Pannu SS. Enhancement of gas-liquid mass transfer using oscillatory flow in a baffled tube. *Chem Eng Sci*. 1993;48(4):799-809. doi:10.1016/0009-2509(93)80145-G.

546606 Design of Vapor Recompressed Dividing-Wall Columns Processing a Light Component Dominant Mixture

When a dividing-wall column (DWC) processes a mixture with a dominant composition of the light component, there exists not only a relatively small temperature elevation from the overhead to the stripping section of its prefractionator but also a large amount of latent heat available in the overhead vapor. These provide opportunities for applying heat pump to the DWC and lead potentially to great improvement in system performance. In this paper, a DWC processing an N-pentane/N-hexane/N-heptane ternary mixture is chosen to examine the characteristics and effect of the applications of heat pumps. We use Stage-Enthalpy Deficit (S-H) profile and Temperature-Enthalpy Deficit (T-H) profile to guide the combinations of heat pumps with the DWC while taking the minimization of total annual cost (TAC) as an objective function. The obtained result shows that multiple heat pumps might be employed to use the compressed overhead vapor to heat the feed, the stripping section of the prefractionator, and the section above the intermediate product of the main column. The first heat pump is employed to preheat the feed, thereby allowing the adoption of feed splitting to strengthen the contact between vapor and liquid phase in the prefractionator. The second heat pump is installed to heat the stripping section of the prefractionator and the third heat pump is installed to heat the separating section above the intermediate product of the main column. The TAC decreases by 25.09% as compared with the DWC. The total utility consumption (TUC) decreases by 62.55% as compared with the DWC. These striking results highlight the great potential of applying heat pump to facilitate the DWC. Current work is underway to simplify the search process in process synthesis and design.

550518 A Modular Design Framework That Reduces the Capital Intensity of Small-Scale Chemical Processes

Modular chemical process systems (MCPS) offer a promising solution for the utilization of many unconventional natural gas resources and chemical feedstocks because of their flexibility, mobility and greater ability to follow time-varying demands. However, many unconventional feedstocks, such as associated gas, stranded shale gas, small natural gas reserves, coalbed methane, landfill gas, biogas and fuel gas, are geographically distributed, show seasonal variability, and are much smaller in volumes than conventional oil and gas reserves with robust supply through well-connected pipelines. At smaller scales, most MCPS technologies currently suffer from high capital intensity. Because of this, modular technologies cannot compete with their large-scale centralized counterparts that benefit from economies-of-scale.

To reduce the capital intensity of small-scale chemical processes, we depart from the classic way of designing single plants in isolation and propose a new methodology that is based on simultaneous design and manufacturing of multiple MCPS with many common functionalities. Instead of customizing equipment for each specific plant, we specify a common design for the equipment that perform the same function across all MCPS. From a manufacturing perspective, designing multiple MCPS simultaneously is advantageous, since it reduces the fixed costs of all common and standardized small-scale modules through economies-of-numbers and lowers the overall capital intensity of each participating MCPS. We present a systematic design framework for modular manufacturing of chemical processes, which is based on a mixed integer nonlinear optimization (MINLP) formulation. The framework explores the common phenomena that occur in different processes and identifies equipment which can be commonly-designed and can therefore be produced in large numbers. The benefits obtained through common functionality-based equipment design are demonstrated through methanol and ammonia production processes. The results indicate that cost reduction in the range 2-9% can be obtained when the inherent commonalities in equipment design for methanol and ammonia processes are leveraged via economies of numbers.

To address variabilities in chemical industries, we apply the concepts rooted in agile design where a multi-product processing train is developed which is capable of operating at different modes with different active process configurations (APCs). This allows us to encounter various seasonal variabilities in terms of feedstock availability and quality. Furthermore, to address uncertainty, we develop a stochastic programming-based two-stage recourse model that obtains optimal plant design plant under uncertain product demands and prices to maximize expected overall profit.

550610 Modeling and Metrics Development for Biomass Pyrolysis Intensification Via Autothermal Operation

The RAPID Center for Process Modeling (CPM) is developing modeling methodology for novel and unconventional processes not covered in commercial process simulators, to facilitate and execute consistent and objective evaluation of performance metrics for RAPID projects. One such process is intensified autothermal pyrolysis of cellulosic biomass. The CPM has identified gaps in the available commercial simulators and developed properties and process models that can be used to model baseline and intensified operations. In this case, the baseline is the current best practice, e.g., entrained-flow noncatalytic fast pyrolysis, while the intensified process is autothermal noncatalytic fast pyrolysis. This presentation includes descriptions of the baseline and intensified process models, the approaches used to model reactive intermediates for which property data are not available, and performance metrics derived from predicted process behavior.

548422 Towards the Identification of Intensified Reaction Conditions Using Response Surface Methodology: A Case Study on 3-Methylpyridine *N*-Oxide Synthesis

Towards the Identification of Intensified Reaction Conditions using Response Surface Methodology: A Case Study on 3-Methylpyridine *N*-oxide Synthesis

Jingyao Wang[†,‡], Yanyan Huang[¡ì,#], Benjamin A. Wilhite[†,‡,], Maria Papadaki[‡,D], M. Sam Mannan[†,‡]

† Artie McFerrin Department of Chemical Engineering, Texas A&M University, 3122 TAMU, College Station, Texas 77843-3122, USA

‡ Mary Kay O¡¯Connor Process Safety Center, Texas A&M University, 3122 TAMU, College Station, Texas 77843-3122, USA

¡ì Department of Chemistry, Texas A&M University, College Station, Texas, USA

CAS Key Laboratory of Analytical Chemistry for Living Biosystems, Institute of Chemistry, Chinese Academy of Sciences, Beijing, 100190 China

D Department of Environmental and Natural Resources Management, School of Engineering, University of Patras, Seferi 2, Agrinio 30100, Greece

ABSTRACT

Identification of inherently safer and intensified reaction conditions is a vital step for transformation of traditional batch/semi-batch synthesis to continuous operation. Speeding up reactions is challenged by several safety and efficiency issues such as thermal runaway risk, side reactions, final product degradation and reactor overpressure. This work demonstrates the use of response surface methodology in search for inherently safer and more efficient intensified reaction conditions for 3-methylpyridine *N*-oxidation reaction performed in a semi-batch pressure-resistant isothermal calorimeter. The experimental conditions are broadly selected to screen various operating variables combinations using Box-Behnken design of experiments. Regression models are developed correlating the catalyst amount, oxidizer dosing rate and reaction temperature with reactor pressure and *N*-oxide yield showing good agreement with experiments in this study and past literature. It is plausible that, even when conducted in a semi-batch mode, the reaction is inherently safer and more efficient under intensified reaction conditions.

Keywords: process intensification, reaction engineering, response surface methodology, reactor safety, semi-batch reactor, inherently safer design

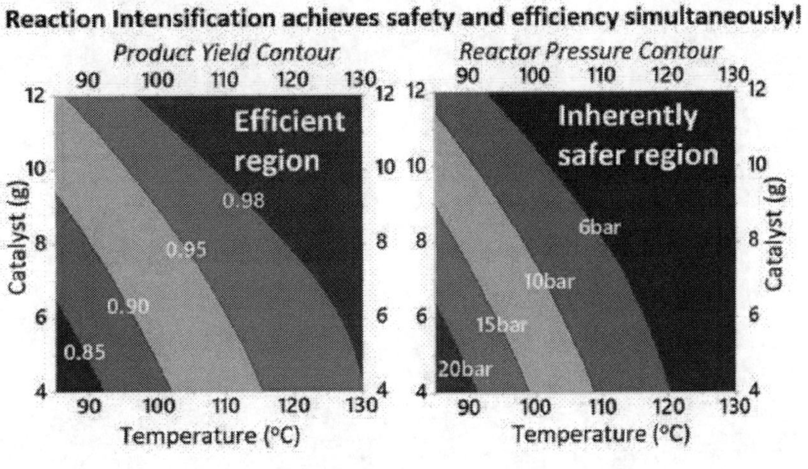

549570 Separation of Dmc/Methanol Azeotrope By Heat Integrated Distillation Equipped with Pervap Membrane

E3Tec has developed Heat Integrated Reactive Distillation (HIRD) process equipped with side reactors and PerVaporization (PerVap) membrane for conversion of CO_2 to dimethyl carbonate (DMC). Separation of DMC/methanol azeotrope for achieving high-concentration (99.9%+) DMC is energy intensive. E3Tec has developed an integrated distillation for separation of DMC/methanol azeotrope by recovery of the overheat latent heat, internal thermal integration and PerVap integrated with the distillation column. The resulting DMC purity of 99.9%+ was achieved as bottom product and DMC in the overhead methanol-rich stream is reduced. Methanol is recycled in the CO_2 conversion process as feed to side reactors; therefore, minimizing DMC content in the recycled methanol stream is preferred for a reversible, equilibrium- controlled reaction. The analysis was based on the ASPEN Plus® simulation of the integrated distillation subsystem. The project was supported by DOE SBIR program.

549973 High-Throughput and High-Efficiency Liquid-Liquid Extraction Column WINTRAY -Review of the Commercial Experiences and Recent Development-

The WINTRAY®, the liquid-liquid extractor, was commercialized to obtain high-throughput high-efficiency and scum-resistant in the liquid-liquid extraction. It has been used in 19 commercial plants worldwide.

For new construction, the size of the column becomes compact by using WINTRAY® because of its high throughput and high efficiency. As compared to extractors coming with driving device, the simple structure which consist of trays and weirs with openings, leads to low capital cost and maintenance cost.

For intensification of existing extraction columns, throughput can be increased by replacing the internals with WINTRAY®. And also, the energy consumption in the solvent recycle section can be saved by reducing the solvent circulation rate of the extraction column.

We herein report the characteristics and commercial examples of process intensification using WINTRAY® . In addition, the result of the further improvement to increase the throughput are introduced.

550485 Material Selection for Packed Columns in the Chlorine Drying Stage of the Chlor-Alkali Process

The Chlor-alkali industry uses electrolytic processes to produce chlorine and caustic soda from aqueous sodium chloride solutions. Chlorine drying is one of the stages of chlorine processing which removes moisture from chlorine by using sulfuric acid. Two or more columns are employed in series for chlorine drying by counter-current contact of gas flow with acid flow in a packed column. For the latter, material selection is pivotal to the process design because even small traces of water in chlorine can be highly corrosive. Water content in chlorine and sulfuric acid drive the material selection, which in turn is dependent on each material's chemical resistance, temperature rating, mechanical strength and cost. Design and economic considerations that lead to selecting materials for the column shell, internals and high capacity random packing are discussed in this paper. Quantitative comparisons of the different packing materials in terms of their hydraulic properties will be shown. This will be supported using select commercial application examples.

549941 Using Ethanol in Continuous Biodiesel Production with Trace Catalyst and CO_2 Co-Solvent

ABSTRACT

The continuous biodiesel production process under sub and super-critical conditions using a trace amount of potassium hydroxide (KOH) as a catalyst have been studied. In addition, CO_2 was added as a co-solvent to reduce the reaction time and increase biodiesel yield. The proposed procedure enables simultaneous transesterification and esterification of triglycerides and free fatty acid (FFA), respectively. The shorter reaction time and milder reaction conditions may reduce energy consumption due to simplification of the separation and purification steps. The process variables, including reaction temperature, ethanol to oil molar ratio, catalyst amount, and process pressure were systematically optimized. The highest biodiesel yield (98.12%) was obtained after 25 minutes reaction time using only 0.11% wt. of KOH and 20:1 ethanol to oil ratio. The process optimum temperature and pressure were 240 °C and 120 bars, respectively. The proposed kinetic model suggested a first order reaction with an activation energy of 19.5 $kJ.mol^{-1}$ and a reaction rate constant of $0.0398/min^{-1}$. The thermodynamic parameter values including enthalpy, entropy and Gibbs free energy for this process were calculated as 15.05 $kJ.mol^{-1}$, -0.27 $kJ.mol^{-1}$ and 153.6 $kJ.mol^{-1}$ at 240 °C, respectively.

AUTHOR INDEX